Reading Essentials
in Science

COMMUNITIES OF LIFE

Rivers

JANE HURWITZ

PERFECTION LEARNING®

Editorial Director:	Susan C. Thies
Editors:	Mary L. Bush, Paula J. Reece
Design Director:	Randy Messer
Book Design:	Brianne Osborn, Emily J. Greazel
Cover Design:	Michael A. Aspengren

A special thanks to the following for his scientific review of the book:
Paul Pistek, Instructor of Biological Sciences, North Iowa Area Community College

---

To Olivia, who brightens our habitat

Image Credits:
© Wayne Lawler; Ecoscene/CORBIS: pp. 4–5; © Yann Arthus-Bertrand/CORBIS: p. 12; © Julia Waterlow; Eye Ubiquitous/CORBIS: p. 16; © Jeremy Horner/CORBIS: pp. 30, 33; © Reuters NewMedia Inc./CORBIS: p. 31; © Vince Streano/CORBIS: pp. 32–33

© Royalty-Free/CORBIS: p. 24; Corel Professional Photos: cover (bottom center, bottom right), pp. 3, 7, 8–9, 10, 13 (top), 14–15, 19, 20–21, 22, 26–27, 35, 38–39, 40; Digital Stock: p. 6; Image Library: p. 23; Perfection Learning Corporation: pp. 9, 11, 13 (bottom), 27, 28, 29; Photos.com: cover (background, bottom left), back cover, pp. 1, 17, 18, 25, 34, 37

For information, contact
Perfection Learning® Corporation
1000 North Second Avenue, P.O. Box 500
Logan, Iowa 51546-0500.
Phone: 1-800-831-4190
Fax: 1-800-543-2745
perfectionlearning.com

1 2 3 4 5 6 BA 08 07 06 05 04 03

ISBN 0-7891-6098-6

Contents

Introduction

If someone asked you to describe the area where you live, what would you say? Do you live in a desert region where it's hot and dry? a forest area with lots of evergreen trees? near a hot, wet tropical rain forest? How would you describe the temperature, sunlight, and rainfall in your hometown? What plants and animals live there?

Biomes

What you are describing is a biome. A **biome** is an **environment** with unique features. For example, an ocean biome has salt water. A **tundra** biome is cold and dry, and often the ground is frozen year-round.

There are many types of biomes, including desert, mountain, tundra, forest, grassland, ocean (saltwater), freshwater, and rain forest. Ecologists have noticed that the same biomes can appear in very different places. Deserts, for example, are found in both hot and cold locations. But even though they are in different parts of the world, all deserts share some characteristics.

Each biome has its own special plant life. Think about the different plants found in a desert, a rain forest, and a grassland. Cacti grow in the desert. Palm trees grow in the rain forest. A variety of grasses cover the grassland.

Biomes are also identified by how plants and animals must **adapt** in order to live there. To live in an ocean biome, plants and animals must be able to live in salt water. In a desert, the wildlife must be able to survive long periods without water. Each biome has its own unique environment to which the plants and animals must adapt.

Ecosystems

Ecologists have also determined that certain groups of plants and animals tend to live together. These groups of living creatures interact with the nonliving parts of the environment, such as rocks or sand. Groups of living creatures that interact with one another and their surroundings are called **ecosystems**.

The red-eyed tree frog, nicknamed "monkey frog" due to its great coordination, lives in the rain forest.

Each biome is made up of many ecosystems. In an ocean, there are different ecosystems living in **coral reefs**, cold Arctic waters, and deep underwater **trenches**. Each layer of a tropical rain forest has its own ecosystems.

Working together, the ecosystems form **communities** of life within each of the biomes.

Roadrunner

Nambi Desert

7

Chapter 1

The Birth of a River

Freshwater

Over 70 percent of the Earth is covered with water. Oceans, lakes, and rivers surround us. However, only 3 percent of this water is fresh. Most of this freshwater is frozen. It is found as ice and snow on mountaintops or at the North and South Poles.

Freshwater is water that doesn't contain salt. When salt water from the oceans **evaporates** into the air, the salt is left behind. When the water falls again as rain or snow, it returns as freshwater.

Freshwater flowing across the Earth has many names. River, stream, creek, brook, runoff, and **arroyo** are some of them. While each of these sources is a part of the freshwater biome, there are differences among them. Some are larger than others. Some are found in different environments or climates. Some move slowly and calmly, while others may roar and rush through an area with great force.

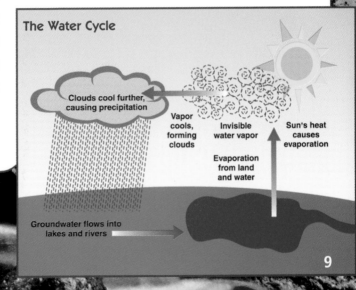

The Water Cycle

Clouds cool further, causing precipitation

Vapor cools, forming clouds

Invisible water vapor

Sun's heat causes evaporation

Evaporation from land and water

Groundwater flows into lakes and rivers

Creeks, streams, and brooks are small bodies of running water. You may have a creek or brook flowing through your backyard or a local park. Most of the time, the water in creeks, streams, and brooks moves slowly. However, heavy rains can cause a temporary increase in this speed as the water rushes toward a larger source of water, such as a river.

Runoff is precipitation (rain or snow) that isn't absorbed into the ground. This water collects on the surface of the Earth and eventually joins a creek, stream, or river.

An arroyo is a stream that flows through a dry environment. When rain is scarce, the arroyo may dry up completely.

All of the smaller freshwater sources eventually lead to a river. A river is a large body of running water that flows into an even larger body of water, such as a lake or ocean. Because of their size, rivers can support a variety of freshwater ecosystems.

A River Begins

Every day, thousands of cars, trucks, and bicycles heading into New York City cross the George Washington Bridge. Flowing more than 200 feet below the bridge is the Hudson River. Where the Hudson passes under the bridge, the river is nearly 5000 feet wide.

But the Hudson River does not start out as a huge body of

A Spanish Stream

Arroyo is a Spanish word meaning "stream or brook."

Arroyo

10

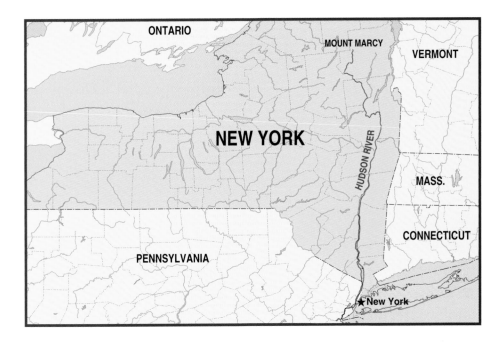

water. The beginning, or **source**, of the Hudson is a small trickle of water, far from cities and cars. The Hudson River's source is a teardrop-shaped lake about the size of a football field. This lake is called Lake Tear of the Clouds. It is located on Mount Marcy, the tallest mountain peak in New York State.

The Hudson leaves the mountains as a small stream. As with all streams, gravity pulls the Hudson downhill toward the ocean. The Hudson flows for hundreds of miles before it reaches the Atlantic Ocean. Along the way, it joins with creeks, streams, and other **tributaries**. Rainwater also seeps into the Hudson and increases the river's size. As the Hudson nears the Atlantic Ocean, it flows past New York City. By this time, many tributaries have joined together to make the Hudson a deep, wide river.

The Hudson's growth from small beginnings is not unusual. All rivers begin from small sources. Since water always flows downhill, rivers and streams usually begin in the mountains or hills. Rainfall, a mountain lake or stream, or a small trickle of water melting from an ancient glacier can form the beginning of a river.

Rivers and streams may also begin from a **spring**. A spring is a source of freshwater flowing out of the ground. Springs form when rainwater or melting snow cannot sink deeply into the earth. This often happens in the mountains where layers of rock prevent water from moving downward into the ground. Instead, the water flows sideways until it springs out of the earth.

Some rivers have a source that is easy to locate. The Ganges River in India starts from the melting water in an ice cave in the Himalayan Mountains. Other rivers, such as the Amazon in South America, have so many tributaries that it is hard to point to a single source.

As a river moves toward its **mouth**, more and more tributaries join with it. Creeks, streams, and smaller rivers flow into the larger river. By the time it empties into a lake or an ocean, a river has come a long way from its small beginnings.

River Systems

A river and all of its tributaries are called a *river system*.

The Mara River in Kenya winds through the countryside.

Human Help

Not all rivers are formed by nature. Some are created by humans. **Canals** are human-made rivers.

Canals are built to lead water from an existing river or stream to a new area. A trench is cut in the ground to make a new path for the water. This new river may be used to carry goods or water crops.

Many large canals were built hundreds of years ago as a way to move heavy goods from place to place. In 1825, the Erie Canal was built in the northeastern United States. It was only 4 feet deep but 40 feet wide. The canal linked Albany, New York, on the Hudson River to Buffalo, New York, on the Great Lakes. The Erie Canal transported goods and immigrants from New York Harbor to towns and other settlements that they could never have reached otherwise.

Some canals run through cities. Examples are Brugge, Amsterdam, and Venice, all cities in Europe.

Canals have been used for watering crops for many centuries. Some canals along the Huang He River in China have been in use for more than 2000 years. Water is moved from the river to the canal by swinging buckets or water wheels. The water is then used by farmers in nearby fields.

Carp in the Canals

When a canal is used for a long time, it develops its own ecosystem. Grass carp live in the canals along the Huang He. These large fish have huge appetites. They feast on the weeds that grow in the river water.

Chapter 2

The Path of a River

Rivers change in size, shape, and direction as they travel from their sources to the oceans. At its source, a river starts out as a trickle of water. Tributaries, snow, and rain add to the flow, making the river larger. As the river grows, the trickle turns to a mass of rushing water. These beginning sections of a river are called the *upper course*.

As the river leaves the mountains and hills, it reaches flatter ground. The speed of the water slows. The path of the river starts to meander, or twist from side to side. From an airplane, a meandering river will look like a giant snake slithering across the land. This section of the river is called the *middle course*.

Closer to the ocean, the land surrounding the river becomes a flat plain. The river is old and near its end. Its speed is slow. This final section of a river is called the *lower course*.

The Upper Course

As a river rushes downward from a mountain, it carves out a thin path, creating a valley. This narrow valley has steep sides like the shape of the letter V. The bottom of the valley is often covered with boulders, rocks, and pebbles.

Rocks and soil are worn away by the water as the river tumbles through the valley. **Minerals** in the rocks dissolve in the water and are carried along in the river's current, or flow. These small pieces of rock, minerals, and soil are called **sediment**. Depending on the type of sediment that it carries, the water in rivers may vary in color. Water in the Tana River in Kenya, Africa, is a dark brown color like tea. The Huang He River in China is a yellow color.

Some sections of the upper course are so noisy that it is hard to hear anything else. Water crashes into large rocks and splashes over waterfalls. These fast-moving sections of the river are called *rapids*. Some fish and insects jump and swim in river rapids.

Huang He means "Yellow River."

The Middle Course

The river valley loses its V shape as the surrounding ground becomes flatter. As more tributaries flow into the valley, the river grows wider. The river has now entered the middle course of its path.

As the middle course meanders from side to side, it picks up soil from the **riverbanks**. The water becomes muddy. The soil that enters the water contains nutrients that feed plants, fish, and other creatures living in the river.

Cities, towns, villages, and farms are common along the middle course of a river. Farmers use the river water for crops and animals. The river flow is usually calm enough for boats that carry food and other products to the towns along the banks. The river also provides a constant supply of freshwater for the communities nearby.

In some parts of the middle course, the land is very flat. At these points, there is almost no riverbank to separate the river from the land. River water mixes with the soil, forming a marsh. A marsh is an area of soft, wet land. Willow trees, reeds, and rushes grow in marshes. Birds, turtles, snakes, and **rodents** make their homes in these moist areas.

Reeds and Rushes

Reeds are tall, skinny grasses with jointed stems. These stems can be used to make instruments.

Rushes are tall plants with round, often hollow, stems. They are used to make baskets, mats, and chair seats.

Rice farmers in China who have run out of room have converted floodplains into farmland.

The Lower Course

By the time a river enters its lower course, it has quieted down. It no longer crashes noisily over boulders or rushes over waterfalls. It doesn't meander. The river valley becomes even flatter than in the middle course. The river is now wide and slow.

The water still carries sediment from the upper river sections. Much of this sediment collects along the bottom of the river valley, forming a deep layer of mud, **nutrients**, and tiny pebbles. This layer is called *silt*.

The land that surrounds the riverbank along the lower course is called a *floodplain*. After heavy rains, the river often floods over its banks. Water and silt spill onto the floodplain. When the water level returns to normal, the silt remains on the floodplain.

The silt left on floodplains provides fresh nutrients for the soil. These nutrients act as fertilizer, creating excellent farmland. Many floodplains around the world are used for growing crops. For example, rice farming is an important activity on the floodplains of the Yangtze River in China.

The river slows down even more as it approaches its mouth. It can no longer carry silt. Near the ocean, the silt drops to the bottom of the river. It collects in mounds called *mudflats* or *sandbanks*. As the silt builds up, new land is created. This land is called a *delta*.

Tons of Silt

Every year, the Mississippi River in North America delivers half a billion tons of silt to the delta at the Gulf of Mexico. This silt makes the delta grow by 295 feet each year.

Along with new land, deltas create new habitats, or homes, for plants and animals. The largest river delta in Africa is located where the Niger River of West Africa empties into the Atlantic Ocean. Large mangrove swamps in the Niger River Delta have made homes for the animals and people there. Elephants and hippopotamuses graze along the delta. Colobus monkeys swing through the trees. Just as in floodplains, the soil in deltas is full of nutrients. Rice farms and oil palm plantations in the Niger River Delta provide food and income for people living nearby.

Mangroves Like a Mixture

Mangrove trees grow where there is both freshwater and salt water. In a delta, the freshwater from the river mixes with the salt water from the ocean. This is the ideal environment for mangrove trees.

Because they are often covered with water or silt, mangroves have roots that can reach above water or mud to breathe. Often the root system of mangroves looks like a tangled mess of drinking straws.

19

Chapter 3

River Habitats

Imagine floating down a river in a canoe. What plants and animals might you expect to see? The river actually provides several different homes, or habitats, for a variety of trees, aquatic (water) plants, mammals, reptiles, birds, and insects.

The Surface

The land along the river, known as the riverbank, is home to many creatures. Tall trees, such as cottonwoods, oaks, elms, and maples, line the banks. Prairie grasses and cattails spread across the land. Violets, orchids, lilies, poppies, chrysanthemums, and many other wildflowers decorate the river's edges.

From the banks, you can watch life on the river's surface. Water striders (also known as *pond skaters*) balance on long, spiderlike legs. These insects can actually run across the water. Dragonflies, mayflies, and caddisflies glide through the air, sometimes dipping down to the water's surface. Air bubbles burst on the top of the water as fish snap at mosquito **larvae**. Lungfish rise to the surface to breathe. Frogs rest on water lilies. Beavers dart in and out of the water, looking for food and building their dens. Muskrats, river otters, and nutrias dive into the water to escape predators.

The Great Egret is over 3 feet tall and has a wingspan of up to 55 inches.

A variety of birds perch, paddle, and wade their way across rivers. Surface-feeding ducks, such as wood ducks or mallards, find food on or near the water's surface. Scaups and pochards are diving ducks. These birds dive underwater for food. Egrets, cranes, and herons are large wading birds. They have long legs and necks that help them wade through rivers in search of food.

Many mammals live near rivers. They use the river as a source of food or water. Deer, wolves, bobcats, raccoons, opossums, and elk can often be spotted at a river's banks.

Playing Possum

Have you ever "played possum" or "played dead"? When opossums are threatened, they play dead. They don't move or blink, and their tongues hang out. They stay "dead" until the predator loses interest.

Some animals move between the banks and the water. Beavers eat aquatic plants that are found on the bottom of the river or the banks. They also climb the banks to eat the bark off trees, shrubs, and plant roots. They build their own homes, or dens, in the river for shelter and protection. Some muskrats build dome-shaped houses on the river, while others dig burrows in the banks. Nutrias are excellent swimmers who find food both in and out of water. A variety of frogs, toads, turtles, and snakes hop, crawl, and slither their way in and out of the water to the banks.

Under the Surface

Fish thrive in the freshwater environment of rivers. Trout swim just under the surface of many cold mountain rivers and streams. Black speckles on the trouts' backs help camouflage, or hide, them as they swim over the pebble-covered **riverbed**. But camouflage isn't always enough to protect the trout. When feeding on mayflies and caddisflies near the surface, trout may be plucked out of the river by hawks or eagles. These birds swoop down and grasp the trout with strong talons, or claws.

In the middle course of a river, the water is often cloudy with sediment. But even if it's hard to see, there is plenty of life just under the surface. In South American rivers, flesh-eating piranhas with razor-sharp teeth eat other fish and animals unlucky enough to come within reach. Tropical, warm-water fish, such as guppies and tetras, swim close to the riverbank where they can hide among water lilies and other plants. Pike, sunfish, perch, and other common fish eaten by humans swarm in rivers in the Northern Hemisphere.

Trout

Catfish swim in the lower course of most rivers of the world. The suckermouth armored catfish of the Amazon River and the red-tailed Nile catfish both live in warm rivers. The blue catfish can live in cooler waters and is common in the Mississippi River of North America.

Bacteria and protozoa are invisible to the human eye but are important to life in the river. These tiny **organisms** live and feed on dead and dying plants in the water. They are the garbage collectors of the river ecosystem.

Bacteria and protozoa are then eaten by worms, insects, and fish. This is an important **food chain** in the river biome.

The Riverbed

The riverbed, or floor, of the river is home to many interesting creatures. Worms, crayfish, and clams are riverbed dwellers. Snails crawl slowly on the muddy floor. Mudpuppies hide under weeds, rocks, and logs.

Crayfish are found in many riverbeds around the world. They eat a variety of plants, snails, insects, and even other crayfish. While crayfish use their claws, or pincers, to hunt along the riverbed, many larger creatures hunt the crayfish. Raccoons and people capture crayfish for food. Otters and turtles also hunt for crayfish.

A crayfish's legs are regenerated, or grown again, if broken off.

Some animals that live on the banks or at the surface of the water dive deep to search for food along the riverbed. Turtles and otters use their powerful legs to make their way down to the riverbed for food. Poisonous water snakes will dive below the surface of the river in search of a frog or newt along the riverbed. Insect larvae on the river bottom provide food for fish.

Beneath the Riverbed

As long as oxygen and water can seep through the rock and silt on a river's floor, creatures can find homes under the riverbed. Salmon bury their eggs in the gravel-covered riverbed of the upper course. In the lower course, crayfish sleep the day away burrowed under the silty riverbed.

Large numbers of insect larvae grow under the riverbed.

Sometimes their protective home of gravel and sand is disturbed by the thin, pointed snout of a soft-shelled turtle as it searches for food. In Australia, the platypus uses its flat, beaklike mouth to sift under the mud and gravel in search of insect larvae.

Mussels bury themselves in the sand and mud of the riverbed. These creatures are also commonly called *shellfish* or *clams*. Tiny plants and animals that float in river water are the main food source for mussels. Hard shells protect them from predators along the riverbed.

If too much sediment washes into a river, mussels and other creatures that live under the riverbed are endangered. As the sediment piles up on the riverbed, oxygen under the riverbed decreases, making breathing difficult.

Chapter 4

Rivers of the World

There are flowing bodies of freshwater in six of the seven continents of the world. Antarctica has freshwater sources, but they are frozen year-round.

Africa

Desert covers much of the northern and southern parts of the African continent. While large rivers do not normally flow through deserts, the Sahara Desert in Africa is home to the longest river in the world—the Nile River.

The Nile has two main tributaries—the Blue Nile and the White Nile. Both tributaries begin at mountain lakes. The Blue Nile carries water that is rich in mud and silt. The water of the White Nile is full of nutrients. The two tributaries meet in the desert country of Sudan.

The drinking water, **irrigation**, and boat travel provided by the Nile allow people to live in the desert land. Farmers dig canals from the riverbank to surrounding fields to provide water for crops. Cotton, corn, and sugarcane thrive along the Nile. These crops can be shipped by boats to other cities and countries along the Nile.

The Congo River is Africa's second-longest river. The Congo travels along a course that crosses the equator twice. It flows through temperate forests, savannas, and tropical rain forests.

About half of Africa is covered in savanna, an area of grassland with a few scattered trees. Herds of giraffes, zebras, elephants, and lions live in the savanna. The Niger River, Africa's third-longest river, flows in a large crescent shape through the savanna of western Africa.

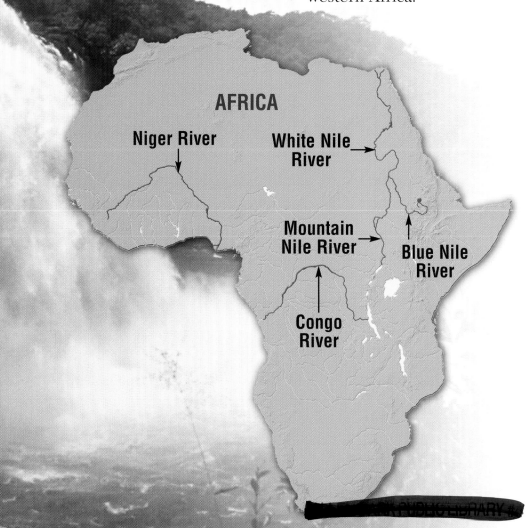

AFRICA

Niger River

White Nile River

Mountain Nile River

Blue Nile River

Congo River

South America

South America is home to the second-longest river in the world. The Amazon River flows through the Amazon Rain Forest, the largest tropical rain forest in the world. Although the Amazon River is not the longest river, it *is* the largest river. Because of the rainy, tropical climate in South America, the Amazon carries more water than any other river. The tributaries that flow into the Amazon cover the entire northern half of South America.

The Carrao River is a smaller South American river. Located in Venezuela, the Carrao flows through the Canaima National Park. The park was created to protect the area from disturbance by humans. Many endangered animal species live in the park, such as the giant anteater, the giant armadillo, and the giant otter.

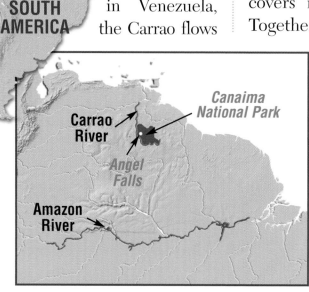

| **Falling Water** |
| Canaima National Park is also home to the tallest waterfall on Earth. Angel Falls was spotted by American pilot Jimmy Angel while he was searching for gold. |

North America

The Missouri-Mississippi river system in North America covers more than 3700 miles. Together these rivers drain water from 31 American states and 2 Canadian provinces.

The Missouri River begins high in the Rocky Mountains of Montana. The youngest portion of the river is less than 3 feet deep. The Missouri travels across several states before joining the Mississippi River north of St. Louis, Missouri.

The Mississippi River begins at Lake Itasca in northern Minnesota. It flows through ten states before emptying into the Gulf of Mexico in Louisiana. Near the Gulf, the Mississippi reaches its deepest depth of 200 feet. The delta near the mouth of the Mississippi provides good land for cotton, rice, and sugarcane farming.

In the northern part of the continent lies the Yukon River. Beginning in Canada, the Yukon flows northwest into Alaska. The Yukon River follows a path just below the **Arctic Circle**, where the winters are long and cold. The river is usually frozen from mid-October until June every year. During the rest of the year, boat traffic flows along the Yukon as it heads to the Bering Sea.

The Colorado River is another large North American river. The lower section of the Colorado River flows through the deserts of the southwestern United States. Like the Nile River, the Colorado provides drinking water, irrigation, and electricity for many desert communities.

Once a mighty, roaring river that carved out the Grand Canyon, the Colorado River is now tamed by ten dams. Dams are barriers made of soil or concrete. The water held behind a dam creates human-made lakes called *reservoirs*.

Asia

The Yangtze River is Asia's longest river. Flowing through China, the Yangtze has over 700 tributaries. Many species of wildlife, such as Tibetan antelopes, Mongolian gazelles, and snow leopards, live along the banks of the Yangtze. Its waters are used for rice and wheat farming as well as transportation for goods and people.

The Ganges River begins in the Himalayan Mountains. It meets with the Brahmaputra River to form the largest river delta in the world. The delta is more than 200 miles of swampland. The rivers finally empty into the Bay of Bengal.

Europe

Despite its small size, Europe has hundreds of rivers crisscrossing its surface. The longest river on the continent is the Volga River in Russia. Even though the Volga is almost entirely frozen for three months every year, over half of Russia's shipping business depends on this river.

River of Life

The Ganges River is a holy place to the many people in India who follow the Hindu religion. The Hindus believe that the flowing water of the Ganges is a symbol that life never ends. Millions of Hindus bathe themselves in the Ganges as part of a religious ritual.

The Danube is another important river in Europe. Wolves, bears, and lynx are found along its banks as it flows through central Europe. The governments of several European countries have agreed to cooperate to protect the river and its surrounding lands from harmful pollution.

River Record

Martin Strel is a long-distance swimmer from Slovenia. In 2000, he swam 1866 miles in the Danube River. By doing this, he won the Guinness World Record for the farthest distance ever swam.

Australia

The rivers of Australia are affected by the continent's dry climate. The longest river, the Darling, has an uneven flow of water. In times of **drought**,

many of the tributaries that feed the Darling dry up. At other times, floods occur when heavy rainfall produces too much water for the tributaries to handle.

Australia's second-longest river, the Murray, has a low flow that also varies according to rainfall. Compared to other rivers of the world, the Murray is a small river. The amount of water that flows through the Murray in a year is about the same amount of water that flows through the Amazon River in one day!

The Murrumbidgee is the third-longest river in Australia. Unlike the Murray and the Darling Rivers, the Murrumbidgee has a fairly even flow of water. This is due to the 14 dams that have been built along the length of the river. Water is let out of the dams as needed.

Three Rivers for Farmland

Together, the Darling, Murray, and Murrumbidgee Rivers provide most of the water that Australians use for irrigating farm crops.

Chapter 5

River Challenges

Whether large or small, dry or overflowing, people have always settled near rivers. Ancient people built their cities near rivers for many reasons. Rivers were a source of drinking water and transportation. Farmers also used rivers and streams for watering their fields.

Rivers and streams are still used today for drinking water, transportation, and farming. But modern people also have new uses for rivers. Electricity is created when the energy in a flowing river is captured by large wheels called *turbines*. Recreational activities, such as waterskiing, fishing, swimming, and sailing also bring people to rivers.

As the number of people in the world increases, the demands on rivers grow too. Supplying the increasing population with clean drinking water becomes more difficult. River habitats are threatened by human actions. Protecting the freshwater river biome has become a challenge we all face.

Pollution

Long ago, when towns first formed along rivers, people used the water as a garbage dump. Human waste and trash were thrown into a river to be washed away. Most rivers were able to survive a small amount of pollution. Animals and plants that lived in the water were able to break down the small amounts of waste that people produced.

But as towns grew, more waste was thrown into rivers. In addition to sewage and trash from households, factories began to dump waste products into rivers. The waste in rivers could no longer be broken down fast enough, so rivers became permanently polluted.

The Ganges River in India flows past cities that are hundreds of years old. Every day, sewage from the millions of people who live along the Ganges is dumped into the river. Besides killing plants and animals in the river, the sewage carries diseases. Industries along the Ganges, such as leather tanneries and cotton dyers, also add poisonous chemicals to the water. Although the Indian government has been trying to clean up the Ganges, it has not made much progress.

It is estimated that over 260,000,000 gallons of waste are dumped into the Ganges River every day.

Ganges River Dolphins Are Disappearing

The Ganges River Dolphin is a long-beaked dolphin that lives in the waters of the Ganges River. These dolphins feed on fish, turtles, and other river creatures. An increase in pollution, boat traffic, dams, and hunting kill many of these dolphins each yèar.

Heat and chemicals from factories can pollute rivers.

Pollution can damage or destroy river habitats. Fertilizers and pesticides used by farmers soak into the ground and can be washed into rivers. These chemicals cause **algae** to grow rapidly. Large populations of algae decrease the amount of oxygen in a river, which harms plants and animals.

Factories use river water in their cooling processes. The water they return to the river is often warmer than the river water. This heat causes changes in the river's underwater habitats.

Chemicals from factories sometimes end up in rivers. These substances can quickly poison many plants and animals. Oil spills in rivers prevent oxygen from entering the water and can kill birds and other animals.

Efforts are being made to clean up river pollution. But these plans are costly and take the cooperation of many people. Some governments have created guidelines to restrict the amount of chemicals and warm water companies can dump into rivers. Some countries have laws preventing river pollution. But until everyone works together, water pollution will continue to threaten people and habitats.

Dams Damage Habitats

Dams also damage river habitats. A dam is a wall or other barrier built to stop the flow of water. Dams are built to provide electricity, drinking water, or irrigation for an area. This may sound like a good thing, but when dams are built, the river and its ecosystems are changed forever.

Bonneville Dam links Oregon and Washington and provides the area with electrical power.

Sediment is not allowed to flow freely through a dam. Plants, animals, and crops downstream from a dam don't receive the nutrients that sediment provides.

Dams also restrict the amount of water that flows below them. Wetlands, such as marshes and swamps, in the lower course of a river may not receive enough water to survive. These areas soak up extra water in river systems that may otherwise cause flooding. Normally, nutrients from sediment and dead plants collect in wetlands. This creates rich soil and food for fish. Dams prevent this from happening, causing permanent damage to the land and the plants and animals that depend on it.

The Future

The electricity that you're using right now may have been produced at a power plant alongside a flowing river. The shower you took last night or this morning might have been provided by a river. The water you drank for lunch could have been river water. You may depend on rivers more than you realize.

The future of rivers is in the hands of people all over the world. Protecting the world's limited freshwater supply will not only allow humans to enjoy its benefits, but it will also save the lives of the millions of plants and animals that live in the river biome.

Internet Connections and Related Reading for Rivers

http://www.newint.org/issue273/ facts.html
Learn about the five largest rivers and how industries threaten their ecosystems.

http://www.factmonster.com/ipka/ A0001779.html
Check out this chart of the major rivers of the world—their sources, outflows, and lengths.

http://www.riversmart.org/ rivers101_ about.cfm#rivers
Take a quick, easy course in "Rivers 101." Read some river facts and follow a river's journey.

http://cgee.hamline.edu/rivers/ Resources/river_profiles/
This site is a quick reference on six great world rivers (Amazon, Zambezi, Volga, Yangtze, Thames, and Mississippi). Statistics, location, climate, and history are given for each river.

http://mbgnet.mobot.org/fresh/rivers/ index.htm
Find out all about various river topics, such as watersheds, how streams become rivers, river zones, river creatures, and more.

....

Endangered Wetland Animals **by Dave Taylor.** Pollution and waste threaten the continued existence of the many species that inhabit the fragile wetlands of this Earth. Crabtree Publishing, 1992. [RL 4 IL 3–7] (4391501 PB 4391502 CC)

Pond and River **by Steve Parker.** An Eyewitness Book on ponds and rivers. Dorling Kindersley, 1988. [RL 6.4 IL 4–9] (5866406 HB)

***The River* by Gary Paulsen.** Because of his success surviving alone in the wilderness for 54 days, 15-year-old Brian is asked to undergo a similar experience to help scientists learn more about the psychology of survival. Dell, 1993. [RL 5.9 IL 4–9] (4400601 PB 4400602 CC)

***A River Ran Wild* by Lynne Cherry.** An environmental history of the Nashua River, from its discovery by Indians through the polluting years of the Industrial Revolution to the ambitious cleanup that revitalized it. Harcourt Brace, 1992. [RL 4.2 IL 2–5] (112201 PB 112206 HB)

***What Is a Biome?* by Bobbie Kalman.** This book introduces biomes, showing and describing the main kinds and discussing their location, climate, and plant and animal life. Crabtree Publishing, 1998. [RL 3 IL 2–5] (5729401 PB 5729402 CC)

•RL = Reading Level
•IL = Interest Level
Perfection Learning's catalog numbers are included for your ordering convenience. PB indicates paperback. CC indicates Cover Craft. HB indicates hardback.

Glossary

adapt (uh DAPT) to learn to successfully live in an environment (see separate entry for *environment*)

algae (AL jee) organisms that grow underwater and float beneath or on the water's surface

Arctic Circle (ARK tic SER kuhl) area at the top of the globe that includes the North Pole, Arctic Ocean, and parts of Europe, Asia, and North America

arroyo (uh ROY yoh) a stream that flows through a dry environment

biome (BEYE ohm) environment with unique features (see separate entry for *environment*)

canal (kuh NAL) a human-made river

community (kuh MYOU nuh tee) organisms that live together in a particular location (see separate entry for *organism*)

coral reef (KOR uhl reef) rocky area in warm, shallow ocean waters created from the remains of animals called *polyps*

drought (drowt) period of time with little or no water

ecosystem (EE koh sis tuhm) group of living creatures that interact with one another and their surroundings

environment (en VEYE er muhnt) set of conditions found in a certain area; surroundings

evaporate (ee VAP or ayt) to change from a liquid to a gas

food chain (food chayn) the order of who eats whom in a community (see separate entry for *community*)

irrigation (ear uh GAY shuhn) watering system for crops

larvae (LAR veye) early, or immature, forms of insects

mineral (MIN er uhl) nonliving material found in nature

mouth (mowth) place where a river empties into a larger body of water

nutrient (NOO tree ent) material that living things need to live and grow

organism (OR guh niz uhm) living thing

riverbank (RIV er bank) rising ground along a river

riverbed (RIV er bed) bottom, or floor, of a river

rodent (ROH duhnt) small gnawing (biting or chewing with teeth) animal, such as a mouse, squirrel, or beaver

sediment (SED uh muhnt) small pieces of rock, minerals, and soil carried by a river

source (sors) beginning of a river

spring (spring) a source of freshwater that flows out of the ground (see separate entry for *source*)

trench (trench) deep canyon, or valley, on the ocean floor

tributary (TRIB you tair ee) small body of water that flows into a river

tundra (TUHN druh) treeless region with soil that is often frozen year-round

Index